Haddar Wafa

Valuation of Tunisian bio-resource in textile finishing

Haddar Wafa

Valuation of Tunisian bio-resource in textile finishing

ScienciaScripts

Imprint

Any brand names and product names mentioned in this book are subject to trademark, brand or patent protection and are trademarks or registered trademarks of their respective holders. The use of brand names, product names, common names, trade names, product descriptions etc. even without a particular marking in this work is in no way to be construed to mean that such names may be regarded as unrestricted in respect of trademark and brand protection legislation and could thus be used by anyone.

Cover image: www.ingimage.com

This book is a translation from the original published under ISBN 978-613-8-47591-0.

Publisher:
Sciencia Scripts
is a trademark of
Dodo Books Indian Ocean Ltd., member of the OmniScriptum S.R.L Publishing group
str. A.Russo 15, of. 61, Chisinau-2068, Republic of Moldova Europe
Printed at: see last page
ISBN: 978-620-4-15460-2

Summary

General introduction

Until the 19th century, chemistry, and in particular organic chemistry, was not yet very developed. The range of dyes used by man was limited to about fifteen, all of which were extracted from natural products. These dyes are often of plant origin: madder root provides red alizarin, Indigofera provides indigo blue, yellow luteolin is derived from gaude, etc.. They are sometimes of animal origin, such as the Phoenician purple, which is extracted from the Murex brandaris shellfish, and the cochineal and kermes reds, which come from insects. The work of man consists then in extracting these dyes from plants or animals. This operation is often long, tedious and provides little colouring matter (12,000 Murex shells are needed to obtain one and a half grams of purple).

In the middle of the 19th century, the emerging chemistry revolutionized the use of dyes. With synthetic dyes and their industrial production, natural dyes became less and less used. It is estimated that nearly 1,000,000 tons of synthetic dyes were used per year (Meena, 2013). However, the use of large amount of synthetic dyes causes environmental pollution because of their often polluted and non-biodegradable waste, which disrupts the ecological balance and causes health risks to human beings.

Thus, and due to the current ecological awareness, the attention of researchers in this field has been shifted to the use of natural dyes for dyeing of textile materials. And in this same context, the present study has been carried out to revive the age-old art of dyeing with natural dyes.

The objective of our work is to demonstrate the possibility of extracting a natural dye from a medicinal plant, *Asphodelus tenuifolius, in* order to dye the wool fibre.

Throughout this manuscript we will present the following parts:

- The first part of our study consists of a bibliographical search in which general information on natural dyeing, the wool fibre, the plant material used, its phenolic compounds, the techniques for extracting natural dyes and the different natural dyeing techniques are successively discussed.

- In the second part, we will present the materials and methods applied in our work in relation to the extraction and dyeing of wool, the colorimetric study and the evaluation of dye solidities.

- The third part will be devoted to the experimental study. During this part we will present the results of the following studies:

• Study of the influence of temperature, pH, time and mass of the plant material on the extraction and quantification of phenolic compounds.

• Study of the influence of these parameters on the quality of wool dyeing.

• Application of the three mordanting methods and study of their impact on the dyeing quality of wool.

We conclude this manuscript by presenting the findings of this unveiled study.

Chapter I: Bibliographic synthesis

Introduction

In this first chapter, we will develop a literature review in which we will focus on the disadvantages of synthetic dyes and the benefits of natural dyes.

We will then mention the different extraction processes and natural dyeing processes. Finally, we will present our plant material *"Asphodelus tenuifolius"*.

I.   General information on dyes

1.   History of dyes

Since the beginning of mankind, dyes have been applied in practically all spheres of our daily life for painting and dyeing paper, skin and clothes, etc. Until the middle of the 19th century, the dyes applied were of natural origin. Inorganic pigments such as manganese oxide, hematite and ink were used. On the other hand, natural organic dyes were applied, especially in the textile industry. These dyes are all aromatic compounds that come mainly from plants such as madder, alizarin, indigo, etc.

The synthetic dye industry was born in 1856 when the English chemist William Henry Perkin, in an attempt to synthesize artificial quinine to treat malaria, obtained the first synthetic dye which he called "mauve" (aniline, a basic dye). Perkin patented his invention and set up a production line, which was later followed by others.

New synthetic dyes began to appear on the market. This process was stimulated by the discovery of the molecular structure of benzene in 1865 by Kekule. As a result, by the beginning of the 20th century, synthetic dyes almost completely supplanted natural dyes (OUBAGHA, 2011).

2.   Toxicity of synthetic dyes

Several studies have shown the toxic effects of synthetic dyes. Their toxicity is in fact due to the content of carcinogenic groups such as aromatic groups, phthalogens, cyanides, barium salt and lead.

These carcinogenic groups (in electrophilic or radical form) attack the pyrimidine bases of DNA and RNA and consequently cause an alteration of the genetic code with mutation and cancer risk (Zollinger, 1991).

In addition, synthetic organic dyes are compounds that cannot be purified by natural biological degradation (Cooper, 1995).

For this reason, dyers have valued the potential of natural dyes and have tried to use them as a promising alternative to synthetic dyes.

3. The advantages of natural dyes

Dyes obtained from natural sources are non-toxic, non-allergic to the skin, non-carcinogenic, easily accessible and renewable. Thus, they have better biodegradability and good environmental compatibility (Ghouilaa, 2011).

In addition, natural dyes are a new source of income as they create a new market and target a specific group of consumers with a luxury product that sells at high prices.

II. Study of the textile support to be dyed: wool

1. General

Wool fibre is a hair secretion from certain animals (sheep, goats, camels). It comes from the shearing of live sheep, especially from Australia, New Zealand, Argentina...

There are 3 main categories of wool:

- Merino wool : It comes from pure breed sheep.

- Crossbred wools : They come from crossbreeding.

- Common wools: They come from indigenous breeds.

2. Structure and morphology of wool

The wool fiber consists of three main elements which are presented by figure 1 (Salpin,2008), they are:

• The cuticle : it accounts for about 10% of the total mass of the wool fibre. Its cells are scale-like on the surface of the fibre. They control the surface properties of the wool fibre.

• The cortex: it represents nearly 85% of the mass of the fibre. Its cells are responsible for the majority of the mechanical characteristics of the fibre.

• The cellular membrane complex: it plays the role of a binder between the cells of the cortex and the cells of the cuticle. Although it accounts for less than 5% of the total mass of a fibre, its influence on the physical and chemical properties of the fibre is important

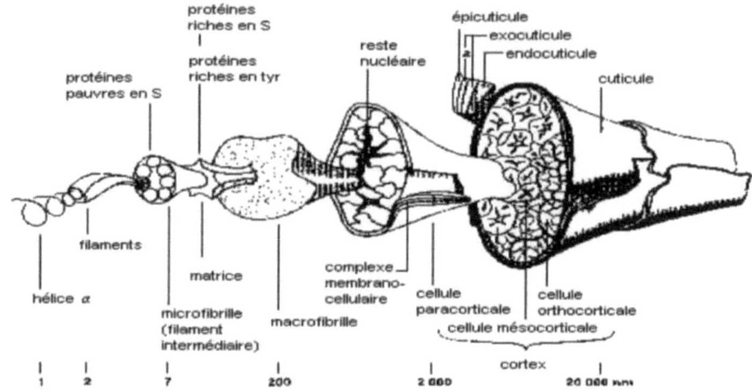

Figure 1: Structure of the wool fibre

3.   Chemical composition of wool

After having explored the morphological structure of wool, we will be interested in its chemical composition, from which chemical reactions will be possible.

Wool is a member of the protein fibre family. It is composed of 97% keratin. The remaining 3% being lipids 2% and mineral salts 1%. It is therefore considered as a fibrous protein (Hamouche, 2012).

Elemental analysis of washed and dried wool gives the following results (Jacques, 2003):

- ✓   Carbon ~50'000

- ✓   Hydrogen ~7

- ✓   Oxygen ~22'%.

- ✓   Nitrogen ~17%.

- ✓   Sulphur ~3.5%.

- ✓   Ash ~0.5'%.

4.   Physical properties of wool

Pure wool has interesting physical characteristics (Hamouche, 2012):

- ➤   The density of the wool is about 20 Kg/m^3.

- ➤   It is a very good thermal insulator which has a very good fire resistance: indeed, pure wool burns without flame from 560°C without releasing toxic gases.

- ➤   It is resistant to combustion because it contains a high amount of nitrogen and sulphur that retards the flames.

- ➤   The thermal conductivity of pure wool is about 0.035 W/mK

➢ Pure wool is highly hygroscopic. This means that it can absorb and release water up to 33% of its weight (1 kg of wool absorbs up to 330 millilitres of water).

➢ Wool is a material that functions as a perfect noise insulator with its ability to attenuate and absorb sound, both at high and low frequencies. The complex structure of wool acts as a wave trap.

➢ The wool fibres have a diameter of 20 to 80 microns and a length of up to 20 cm.

➢ Wool is easy to keep clean: it is a low static generator, so it does not retain dust and does not get dirty.

5. Chemical properties of wool

We will mention some chemical properties of wool:

➢ Action of the heat : We use the heat either to dry the wool, or for certain treatments: calendering, pressing, decatising...

• At 100°C, the wool becomes anhydrous and plastic (a property applied to pressing and calendering). It does not alter and retains the property of recovering its natural moisture on cooling.

• At 130°C, the wool begins to decompose, releasing sulphurous ammonia compounds; this alteration results in a yellowing that is difficult to correct.

➢ Action of water: Cold water has no action, but hot water facilitates the felting of the fibres and prolonged boiling attacks the wool. Humidity acts unfavourably on wool, although water is chemically inactive, facilitating the formation of moulds, often invisible to the naked eye, which eventually destroy the fibres.

➢ Action of acids: Dilute acids have practically no effect on wool fibres. On the other hand, concentrated acids destroy the wool.

• The prolonged action of dilute sulphuric acid promotes felting of the wool to the same degree as soap.

• Hydrochloric acid, liquid or gaseous, acts in much the same way as sulphuric acid; it is not used in dyeing but is frequently used in carbonizing.

• Sulphurous acid is widely used as a bleaching agent for wool.

➢ Action of oxidants : Oxidants generally attack the wool, bleach dissolves the fibre (at 11° Cl specific solvent of the wool).

➢ Action of the reducers: The bleaching of wool can be carried out by reduction.

When the wool is immersed in an aqueous medium, it may be as depicted in Figure 2, depending on the pH conditions, where "W" represents the fiber:

$$NH_3^+ - W - COOH \rightleftarrows NH_3^+ - W - COO^- \rightleftarrows NH_2 - W - COO^-$$

Acidic conditions   Conditions   Basic conditions
                           Isoelectric

Figure 2: Schematic of a wool fibre under different pH conditions

According to the literature, the isoelectric point of wool, an amphoteric material due to its composition of amino acids with acidic and basic characteristics, is around 4.2 (Jocic et al., 2005).

III. Extraction processes for natural dyes

1. General

The extraction of coloured species is very old. It was already done in prehistoric times in order to extract from plants and certain animal organs, food, pharmaceutical or odoriferous products, in the form of beverages, drugs or perfumes

In the extraction and separation processes of specific molecules (active molecules) present in a solid medium, the operation often involves, from a technological point of view, the diffusion within the solid of a fluid (liquid) carrier, known as the extraction solvent; the extraction is thus presented as a solid-liquid interaction

In our work, we are interested in extraction processes from a solid system, the plant. There are several techniques for extracting high value-added products from plants. Let us note at this level that there are classical extraction methods and others that are alternative.

2. Conventional methods of extraction

2.1 . Solvent extraction

Solvent extraction consists of passing the substance to be extracted into a solvent by solubilization. This can be water, but generally it is an organic solvent from petroleum chemistry: cyclohexane, petroleum ether, toluene... Solubilization can be carried out by different methods.

• **Percolation:** this consists of letting a solvent (usually very hot) flow over a bed of finely divided solids. The preparation of coffee is based on this operation.

• **The decoction: it is** the operation in which the solid is plunged into the liquid solvent boiled. It is a brutal operation that should be reserved for the extraction of non-thermolabile active principles. It is however very fast and sometimes essential.

• **Infusion:** this is a decoction during which the solvent is heated without boiling, followed by the cooling of the mixture.

- **Maceration:** it is an infusion in a cold solvent. To be effective, a maceration can last from 4 to 10 days.

- **Digestion:** it is a hot maceration. This operation and maceration are used particularly in pharmacy and perfumery.

- **Elution:** it consists in removing a solute fixed on the surface of a solid by simple contact with a solvent. It is frequently used in analytical methods (Ben Amor, 2008).

2.2 Hydrodistillation

Steam distillation, also known as hydrodistillation, is a widely used technique for the extraction of essential oils. The advantage of this technique lies in the lowering of the distillation temperature; the compounds are therefore entrained at temperatures much lower than their boiling point, which avoids their decomposition.

Figure 3 shows the hydrodistillation device.

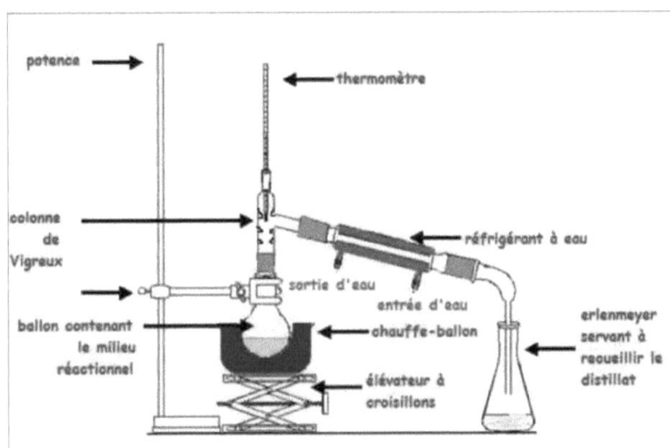

Figure 3: Hydrodistillation apparatus

2.3 The Soxhelt

Soxhlet extraction is a general and well-established technique, and outperforms other conventional extraction techniques, except in the case of extraction of thermolabile compounds.

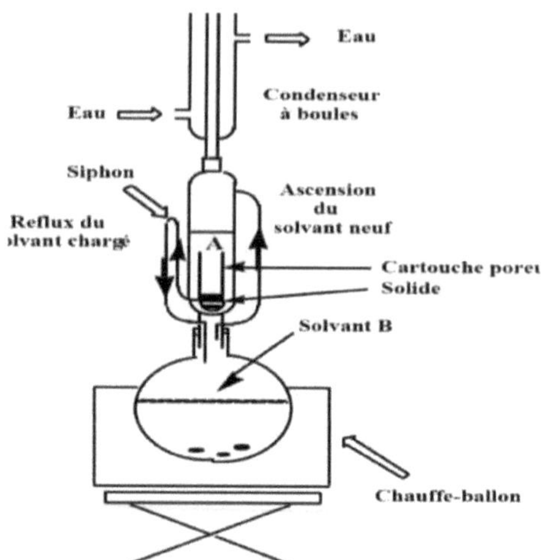

Figure 4: Experimental setup of a Soxhlet extractor

In a conventional Soxhlet system as shown in Figure 4, plant material is placed in a cartridge, and filled with fresh solvent condensed from a distillation flask. When the liquid reaches the overflow level, a siphon sucks the solution from the cartridge and discharges it back into the distillation flask, carrying the extracted dissolved material into the bulk liquid. In the flask, the dissolved material (solute) is separated from the solvent by distillation. The solute remains in the flask and the fresh solvent flows back into the solid bed. The operation is repeated until complete extraction is achieved.

The diverse applications, good reproducibility, efficiency and ease of handling of the extracts are the specific advantages of Soxhlet extraction. On the other hand, the large amount of solvent as well as the long duration of the operation have led to wide criticism of this method. Indeed, the possibility of thermal degradation of the target compounds cannot be ignored since the extraction usually takes place at the boiling point of the solvent during a rather long time (Amiot-Carlin, 2004).

3. Alternative methods

3.1 Ultrasonic assisted extraction

Above 20 kHz, sound waves generate mechanical vibrations in a solid, liquid or gas. Unlike electromagnetic waves, sound waves can propagate through a material and involve cycles of expansion and compression as they propagate through the medium. Expansion can create bubbles that form, expand and collapse in a liquid. Near a solid surface, the cavity collapse is asymmetric and produces a high velocity liquid jet. The liquid jet has a strong impact on the solid surface (Amiot-Carlin, 2004). The mechanical effects of ultrasound induce a greater penetration of the solvent into the cellular materials and improve the mass transfer.

The most significant advantages of ultrasonic extraction are related to the increase in extraction yield and an acceleration of the kinetics compared to a conventional extraction. It allows to work at relatively low temperatures and to avoid the thermodestruction of the compounds. This method does not allow the solvent to be renewed during the process. The limiting step is the filtration and rinsing after the extraction.

3.2 Microwave assisted extraction

Microwaves are electromagnetic radiation with frequencies ranging from 0.3 to 300 GHz. Microwaves can penetrate biological materials and act on polar molecules such as water to impart a fluctuating motion, thus increasing the temperature of the material in question at the depth of penetration.

Microwave-assisted extraction offers rapid energy transfer and simultaneous heating of the solvent and solid plant matrix. By absorbing the microwave energy, the water in the plant matrix promotes cell disruption, facilitating the release of chemicals from the matrix and improving extraction (Kaufmann, 2002). This method uses smaller amounts of solvent, is inexpensive and is considerably faster. However, the operating temperature of this technique is relatively high (100-105°C), which poses problems when it comes to the extraction of antioxidants.

3.3 Extraction with supercritical fluids

The supercritical state of a fluid is achieved by driving the gas above its critical temperature or compressing the liquid above its critical pressure. The critical temperature is the temperature above which the liquid phase of the substance cannot exist, regardless of pressure; the vapour pressure at the critical temperature is the critical pressure (Sihvonen et al. 1999).

The advantages of this method over conventional methods are based on a shorter extraction time, high selectivity and the ease of removing the solvent after extraction by simple decompression (Danielski, 2006).

3.4 Extraction in batch mode

It is a simple device consisting of 500 ml flasks (where the raw material and solvent are placed) and a shaker (Figure 5) (PENCHEV, 2010).

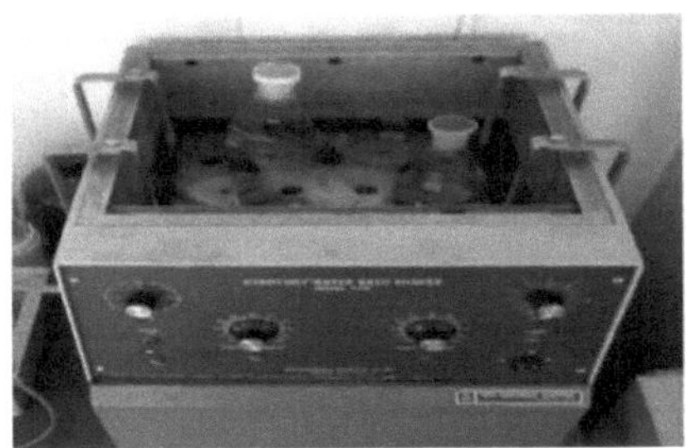

The major advantage of batch extraction by agitation compared to soxhelt is the possibility of working easily with mixtures of solvents (e.g. various alcohol-water ratios) and of controlling the extraction temperature, avoiding the risk of destruction of thermolabile compounds. It is a simple and efficient method but requires relatively long subsequent filtration and concentration procedures (Grigonis et al. ,2005).

3.5   Accelerated solvent extraction

Accelerated solvent extraction is a solid-liquid extraction process performed at elevated temperatures, usually between 50 and 200°C and at pressures between 10 and 15 MPa. The extraction is carried out under pressure to maintain the solvent in its liquid state at elevated temperature. The solvent is always below its critical state during accelerated solvent extraction. The elevated temperature accelerates the extraction kinetics and the elevated pressure maintains the solvent in its liquid state, thus achieving an efficient and rapid extraction. Compared to traditional Soxhlet extraction, accelerated solvent extraction involves a large decrease in the amount of solvent and extraction time.

Accelerated solvent extraction is usually used for the extraction of stable organic pollutants at high temperatures from environmental matrices. Very few applications of accelerated solvent extraction have been reported in the field of plant molecules. Accelerated solvent extraction has been developed for the rapid extraction of cocaine and benzoylecgonine from coca leaves using methanol as solvent (Brachet et al., 2001).

IV.   Natural dyeing processes

There are three natural dyeing processes that have been recorded since ancient times (Mansour, 2005).

- ➢   Cold dyeing
- ➢   Direct dyeing (without mordanting)
- ➢   Dyeing with mordanting

1.   Cold dyeing

It is also known as "Dyeing with fermentation". Specialists believe that this is the oldest form of dyeing in the Middle East, a geographical area where the climate allows such an application (average temperature high enough to cause fermentation). The fabric is immersed in a dye bath and left to ferment at room temperature for a period that can exceed two weeks.

2.   Direct dyeing

It consists in obtaining a dye bath by simply macerating or boiling certain plants in water. The fibres to be dyed are then immersed in the resulting bath. This dyeing technique is applied to dyes whose molecules have groups that are favourably disposed in relation to those of the fibres to be dyed. Chemical bonds can therefore be formed between the pigment molecule and the fibre.

3. Dyeing with mordanting

Since the majority of dyes extracted from plants cannot form solid bonds with the textile fibre, it was necessary to add metallic salts called "mordants". The mordants allow the dyes to be fixed and different shades to be obtained.

3.1. Examples of biting materials used

To reinforce the colours and the solidity of the dyes, the dyers tried very early and then adopted the use of various mordants:

> organic substances (tannins, plant juices and decoctions)

> mineral substances (crystals, sludge and metal oxides)

We will present some examples of the different metal salts used as mordants.

- *Tin etching*

The most commonly used tin salt for etching is stannous chloride $SnCl2$. It comes in the form of white crystals. It allows to obtain very bright tones, and is used especially for red and orange stains. It must be used with care because it can damage certain textiles such as wool.

- *Mordanting with alum alone or with alum and cream of tartar*

These are the two mordants that modify the colour of the dye the least.

■ Alum with the formula $KAl(SO4)2,12H2O$ is the most universally used mordant. It comes in the form of small transparent crystals.

■ Tartar is the salt crust that settles in wine barrels. It is used in the form of potassium bitartrate or cream of tartar. The addition of cream of tartar during the mordanting process increases the amount of dye fixed on the textile.

- *Iron etching*

Iron salts, unlike aluminium mordants (alum), alter the natural colour of dyes, making them darker. They react with the tannins of certain plants to give a colour in the black tones. The most commonly used iron mordants are ferrous sulphate ($FeSO4, 7H2O$) and iron acetate.

- *Copper mordanting*

Like iron mordants, copper mordants change the colour of natural dyes by darkening them. For example, they will turn yellow dyes to olive green or bronze green. They are used in the form of copper sulphate and copper acetate. Copper acetate comes in the form of emerald green crystals while copper sulphate comes in the form of blue crystals.

4.2. Mordanting methods

Mordant can be added at different stages of the dyeing process. In fact, there are three methods of mordanting:

> ➢ Pre-staining (staining before dyeing in a separate bath)

This is the most common method used in European industrial dyeing since the Middle Ages. It allows the dye bath to remain unaltered and several fabric samples to be prepared in advance.

> ➢ Simultaneous etching

This method is quick. It allows to obtain satisfactory results with wool, in particular if one wants to obtain light colours. The dye is first extracted by decoction if the pigment is not available in powder form. The mordant is added and the fibre is added after it has dissolved completely.

> ➢ Post mordanting (mordanting at the end of the dyeing process)

This method allows to intensify, to fix even more solidly and to shade the obtained dye. It is often applied in addition to a mordanting before dyeing.

V. **Study of plant material:** *"Asphodelus tenuifolius*

1. Botanical description

- **Asphodelus: is** the generic name of Asphodel indicating in Latin and Greek several species of lily, dedicated to the gods of hell and death who were supposed to eat tubers

- **Tenuifolius** : it means a plant with narrow leaves

It is an annual plant, with tubers at the base, growing from 20 to 30 cm in height, slightly fleshy, cylindrical, hollow in the middle, glabrous and dark green. The leaves form a circular set at the base of the plant. The inflorescence is a simple, elongated, indeterminate cluster with stemmed flowers. The small flowers have six white tepals. The fruits are small capsules with minute black seeds. Flowering takes place in early spring from March to May (Batanouny, 1999). Figure 6 represents our study plant *"Asphodelus tenuifolius"*.

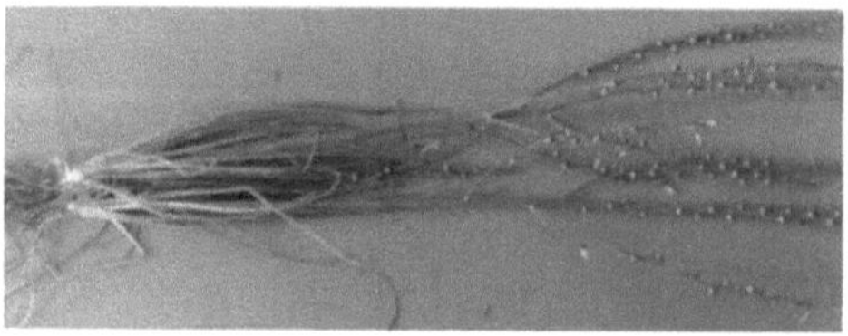

Figure 6: Asphodelus tenuifolius

2. Geographical distribution of the species " *Asphodelus tenuifolius* ".

The species " Asphodelus tenuifolius " grows in arid pastures and steppes. In TUNISIA it is widespread in Djerid, Tataouine, Ben Gherden, Gafsa, Gabes, Sfax, Kairouane Maknassy, El Djem and Ain Cherichera. On the world scale it is distributed from Tripolitania to Spain, Greece, Palestine, Sahara.

3. Ethnobotanical study

"*Asphodelus tenuifolius*" is widely used in traditional medicine by the population of the Ghardaïa region (Algerian northern Sahara) in herbal tea, powder and ointment for the treatment of fevers, indigestion, constipation and skin lesions.

In Egypt, Algeria and Morocco, the seeds are reported to be diuretic and are consumed with yogurt. It is also used for wound healing. The leaves are fried or boiled and are sometimes put in the sauce for couscous. The roots are used in fumigations against jaundice, dried and used as dressings against ulcers and reduced to ashes, to serve as a diuretic (Bellakhdar, 1997).

4. Polyphenols

4.1. Definition of polyphenols

Polyphenols are molecules synthesized by plants during secondary metabolism to defend themselves against environmental aggressions. They are located in different parts of plants depending on the plant species and the polyphenolic group considered. These compounds include a multitude of molecules and represent one of the most important groups present in the plant kingdom. As a definition, we can say that polyphenols are water-soluble organic compounds, with molecular weights between 500 and 3000 Dalton, and having, in addition to the usual properties of phenols, the ability to precipitate alkaloids, gelatin and other proteins (Akroum, 2011).

4.2. The different classes of polyphenols

Polyphenols have several phenolic groups with or without other functions (alcoholic, carboxyl...). In this family of molecules, there are many substances, which can be classified according to their structure in five main groups (Akroum, 2011).

a.   Phenolic acids

Phenol acids, or phenolic acids, have one acid function and several phenol functions. They are colourless and rather rare in nature.

b.   Flavonoïds

Flavonoids have a basic skeleton formed by two C6 rings (A and B) linked together by a C3 chain that can evolve into a heterocycle (ring C) (Figure 3). They give colours ranging from light yellow to golden yellow. According to structural details flavonoids are divided into 6 groups: flavones, flavonols, flavonones, isoflavones, chalcones, aurones.

c.   Anthocyanins

Anthocyanins give very varied colours: blue, red, mauve, pink or red. These molecules have, like flavonoids, a basic skeleton in C15 formed by two rings A and B, and a heterocycle (ring C), but their main characteristic is that the latter is positively charged.

d.   Flavours

Flavans are in the form of monomers (eg: catechin) or in the form of polymers (dimers, trimers ... of catechin). They exist in the form of several stereoisomers from two asymmetric carbons: C2 and C3. Flavans are very common in plant bark

e.   Tannins

Tannins are macromolecules that can be divided according to their structure into two main groups:

>   Hydrolysable tannins: are gallic acid esters that bind to glucose molecules.

>   Condensed tannins: proanthocyanidins: these are heterogeneous phenolic compounds. They are found in the form of oligomers or polymers of flavan-3-ols, 5-desoxy-3-flavonols and flavan-3,4-diols.

>   The polymers give a structure bristling with phenolic OHs capable of forming stable bonds with proteins.

4.3. Polyphenols from *Asphodelus* tenuifolius

Solid/liquid chromatographic studies at atmospheric pressure and solid/liquid high performance chromatography (HPLC) allowed the pure isolation of five natural substances from the butanolic extract of ***Asphodelus tenuifolius*** species which are presented in Table 1 (Faidi Khaled, 2014).

From this table, we can see that the polyphenols contained in the plant material studied are mainly flavonoids.

Table 1: Polyphenols isolated from Asphodelus tenuifolius

Structure	Name	Description
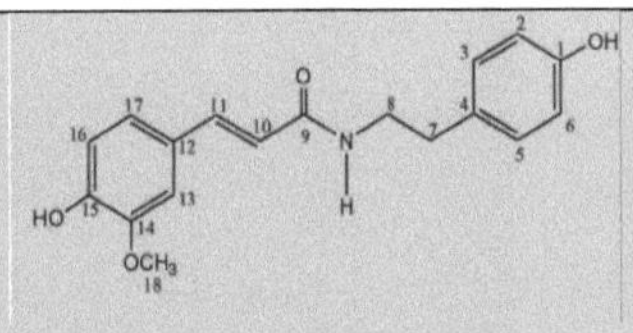	Trans-N-feruloyltyramine	It is a phenolic derivative
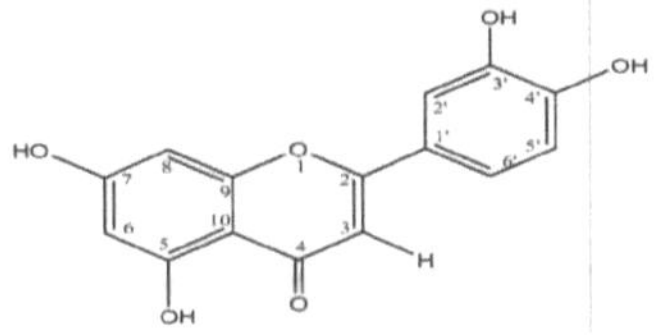	Luteolin	Belongs to the flavonoid family. It is one of the most common flavones. It was first isolated from gaude, a plant then used to make yellow dye, by the French chemist Michel-Eugène Chevreul.
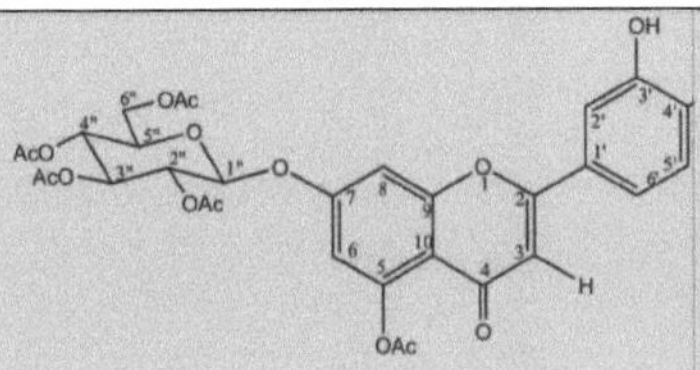	Luteolin 7-0-p-D- Glycopyranoside	Belongs to the flavonoid family.
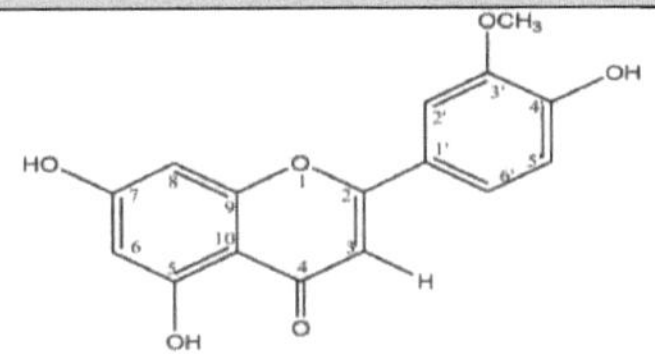	Chrysoeriol	A compound of the flavone family.
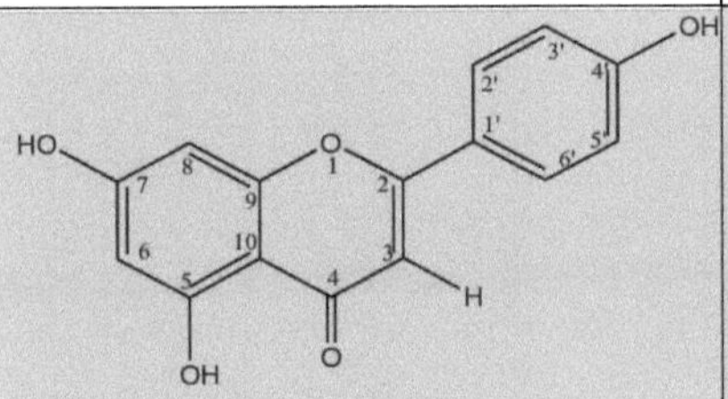	Apigenine	A chemical compound of the flavone family, a subclass of flavonoids, which has anti-inflammatory properties.

5. Compounds responsible for the colour of *Asphodelus tenuifolius*

These are essentially flavonoids. These are natural compounds that belong to the polyphenol family.

They are responsible for the colouring of plants (in addition to chlorophyll, carotenoids and betalains).

Flavonoids have a basic skeleton formed by two C6 rings (A and B) linked together by a C3 chain that can evolve into a heterocycle (ring C). They give colors ranging from light yellow to golden yellow (Zeghad, 2009).

VI. Evaluation of the dyeing quality of dyed textile supports

Colorimetry is the science of visual equivalence related to color perception. A colorimetry system is able to tell whether two colors are equivalent or different by a number of characteristics.

The CIE (International Commission on Illumination) has standardized a "reference observer" to formalize the equivalence. This observer is valid for specific conditions of visual angle of observation and adaptation.

Refinements for different observation conditions are made possible by the evolution of knowledge, both from the point of view of neurophysiology and the psychology of vision. In particular, difference colorimetry, dealing with the problem of color metrics, is making a wide appearance in the scientific world.

1.   Measurement of color coordinates

The objective of this test is to determine for each dyed sample its colorimetric coordinates (L* a* b*) or (L* C* h*) in CIELAB space, and then evaluate the variation of each coordinate.

In this space a color can be defined in two ways:

- By its cylindrical coordinates: L*, a*, b*.

- By its polar coordinates: L*, C*, h*.

Designation of coordinates in the CIELAB space :

- The luminance L* represents the brightness (between 0 and 100).

- The a* component represents the green-red color opposition.

- The b* component represents the blue-yellow color opposition.

- The C* component represents the saturation (distance from the center).

The h* component represents the angle of the hue (between 0 and 360°).

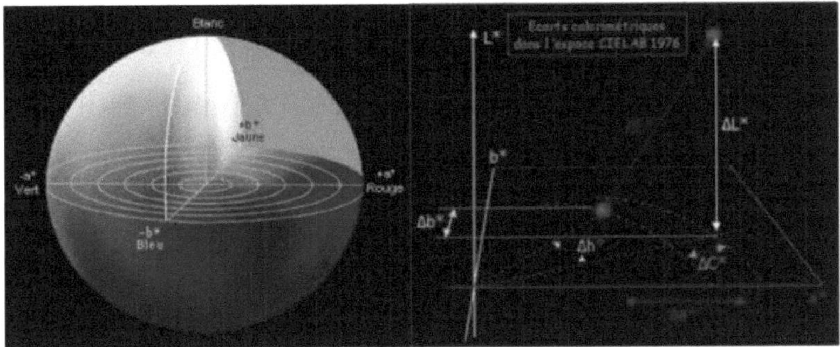

Figure 7: Representation of the CIELAB colour space

2.   Measurement of the degree of absorption (K/S)

The dyeing yield is evaluated by calculating the degree of absorption (K/S) of the dyed samples. It is a parameter that can be estimated by determining the reflectance R of the dyed sample in relation to the absolute white at a specific wavelength. Kubelka and Munk (kubelka and munk, 1948) proposed the following formula to calculate the degree of absorption (K/S):

$$K/S = (1 - R)^2/2R - (1 - R_0)^2/2R_0$$

Where: S is the light scattering factor or diffusion coefficient, K is the absorption coefficient, R is the reflectance of the dyed wool sample and R_0 is the reflectance value of the undyed wool sample (before dyeing).

3.   Measurement of dye solidities

By definition, the fastness of dyes is the ability of the dyes fixed to the fibre to resist destructive agents either during manufacture or during their subsequent use.

The test methods are grouped into two classes:

✓   Solidity of use (light, washing...)

✓   Manufacturing solidities (chlorine, decatizing.)

Each strength test is performed on a sample prepared under specific conditions. After the test, we determine the value of the strength by one or two ratings: colour degradation and bleeding on white control fabrics.

Strength tests are performed under specific conditions that are detailed in well recognized standards. Typically, testing consists of subjecting the sample to the desired test and then evaluating the change in grade through use:

✓   The gray scales: they are presented in the form of two scales, each of which consists of 5 levels. One to determine the degree of white and the other to determine the shade degradation of a dyed sample.

✓   The blue scale: it is composed of 8 wool strips dyed in blue with different types of dyes. It is used only to evaluate the light fastness. The light fastness index is a number Conclusion

This literature review has identified the possibility of exploiting our medicinal plant for textile uses to produce a new range of environmentally friendly dyes.

Conclusion

The present bibliographical study has made it possible to present the information relating to our study, namely, the extraction techniques commonly used as well as the natural dyeing techniques. We have also presented the plant material *"Asphodelus tenuifolius"*, *the* main object of our study, as well as the textile support to be dyed. Finally, we concluded with an overview of the evaluation of the dyeing quality of the dyed textile supports.

Chapter II: Materials and Methods

Introduction

This second chapter presents the principles and equipment of the different techniques used in the experimental processes, as well as the parameters used in this work.

Firstly, we present the characteristics of the textile support that is the subject of this study. Then, we describe the equipment and methods used during the extraction, dyeing and mordanting. Finally, we present the analytical methods chosen during the experimental part.

I. Characteristics of the textile support used

During this study, we chose a woolen fabric as a textile support, whose characteristics are presented in the following table.

Table 2: Wool support characteristics

FEATURES	VALUE	STANDARD USED
Armor	Canvas	NFG-07-154
Grammage	132 g/m^2	ISO-3801
Density Frame	14 duites/cm	ISO-7211-Part II
Density Chain	14 threads/cm	ISO-7211-Part II

II. Experimental protocols used

1. Pre-treatment of the textile substrate

In order to increase the degree of whiteness, we have carried out a bleaching treatment with hydrogen peroxide for the wool fabric.

1.1 Materials

To perform the wool bleaching, we used a 10-4 g precision balance and a hot plate.

1.2. Method

Wool bleaching recipe:

Bath :

- $RoB = 1/40$

- Hydrogen peroxide (H_2O_2): 25 ml/l

- Ammonia

Process:

- Bleaching temperature = 50°C.

- Bleaching time = 60 minutes.

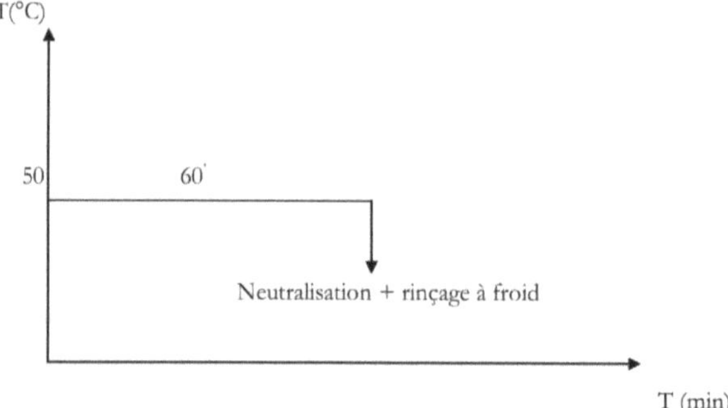

Figure 8: Wool bleaching process

After bleaching, we carry out a bath neutralization in order to avoid the degradation of the wool during drying.

❖ Recipe for neutralizing wool:

Bath :

- RoB = 1/40

- Formic acid 1 ml/l

Process:

- Neutralization temperature = 25°C.

2. Extraction process

2.1. Material

The extraction is carried out from the plant "*Asphodelus tenuifolius*" in a pressure autoclave type AHIBA nuance (Data Color International, USA) under agitation.

a.   Plant material

The plant material consists of the aerial part of *Asphodelus tenuifolius*. The leaves were collected in March 2012 from the region of Kairouan. After drying at an ambient temperature and protected from sunlight, in order to preserve the integrity of the molecules as much as possible, the plant material was coarsely ground in a traditional grinder.

b.   Laboratory scale dyeing machine (AHIBA nuance top speed II)

It is an autoclave under pressure. It has a programmable speed allowing to dye all types of fibres up to a temperature of 140°C. The solutions of dyes and auxiliary products are manually dosed for 12 bottles of dye, heated by 4 infrared emitters. The actual temperature of the liquid is measured directly by a control probe. The units of the AHIBA nuance Top speed II dyeing machine continuously agitate the liquid and the material. The speed of rotation can be selected between 5 and 50 rpm. The direction of rotation is programmable and can be changed. These systems have a memory capacity of 30 programs, each with a maximum of 15 steps, such as temperature, gradient and holding time. Figure 9 shows the AHIBA machine.

Figure 9: AHIBA speed II

2.2. Methods

We soak a certain mass of our already ground plant in 100 ml of water. The extraction is carried out using AHIBA speed grade II. The extraction process is shown in figure 10. At the end of the extraction process, we filter our extract by removing the plant material. We obtain a liquid extract that will be used to make the dye.

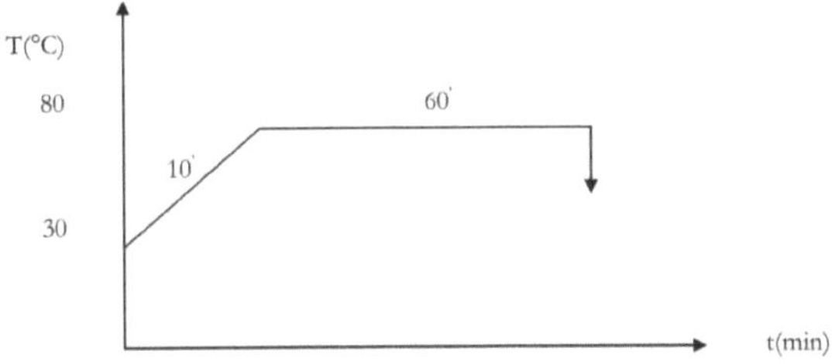

Figure 10: Thermal curve of the extraction

3. Dyeing process

3.1. Material

The dyeing is carried out in the autoclave under pressure and with agitation according to the thermal curve shown in figure 9.

3.2. Method

The dyeing technique we used consists of impregnating the textile material in the aqueous extract of *Asphodelus tenuifolius* prepared under the best conditions (T=40°C, Mass= 4g, time= 60 min) in the AHIBA under pressure. Figure 11 shows the thermal dyeing curve.

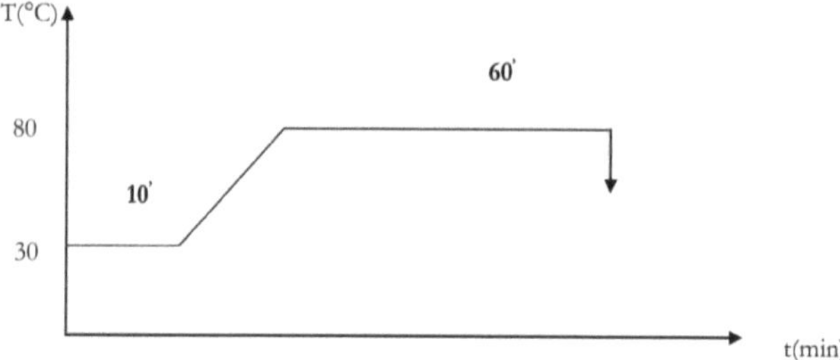

Figure 11: Thermal dyeing curve

4. Etching processes

4.1. Dyeing with pre-mordanting

This technique was applied using a RoB equal to 1/40 and putting a quantity of mordant equal to 3% of the mass of the sample to be dyed. The mordanting process is as shown in Figure 12.

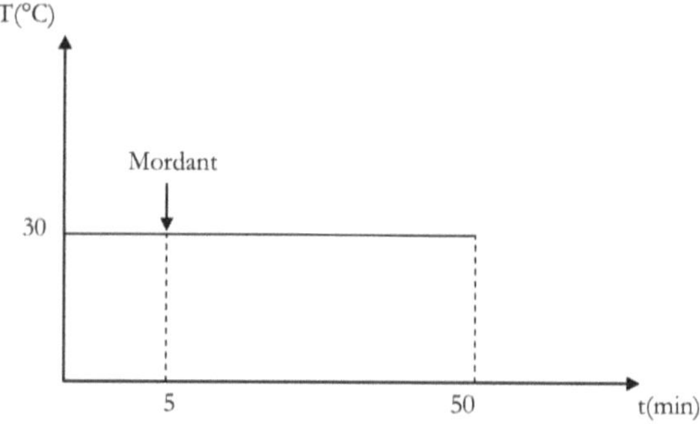

Figure 12: Thermal curve of pre-ordering

After this etching step, we did a simple rinse with cold water. Then, we performed the dyeing in the aqueous extract of the margine according to the technique described in figure 9.

Then we did a rinse under the following conditions:

- RoB = 1/40

- non-ionic detergent: 3g/l.

- Rinse time= 2 minutes.

- rinsing temperature is equal to the ambient temperature

Finally, rinse with cold water and dry.

4.2. Dyeing with simultaneous mordanting

This type of mordanting consists in putting simultaneously and from the beginning of the dyeing process the sample and the mordant in the aqueous extract of the margine according to the technique described in figure 10.

Then we have a soaping under the same conditions already described for the pre-treatment.

4.3. Dyeing with post-etching

After staining the sample according to the previous technique (figure 9), we introduced the sample into the etching bath and followed the same process described above. Then, we have a rinse identical to the one described for the other two types of etching. Similarly, we did a soaping with cold water followed by a drying.

5. Evaluation techniques used

5.1. Colorimetric evaluation

a. Material

To carry out the colorimetric evaluation, we used the Spectraflash spectrocolorimeter, which is shown in Figure 13.

In fact, the dyeing yield is evaluated by calculating the degree of absorption (K/S) of the dyed samples. It is a parameter that can be estimated by determining the reflectance R of the dyed sample in relation to the absolute white at a specific wavelength. Kubelka and Munk (Kubelka, 1948) proposed the following formula to calculate the degree of absorption (K/S):

$$K/S = \frac{(1-R)^2}{2R} - \frac{(1-R_0)^2}{2R_0}$$

Where: S is the light scattering factor or diffusion coefficient, K is the absorption coefficient, R is the reflectance of the dyed wool sample and R_0 is the reflectance value of the undyed wool sample (before dyeing).

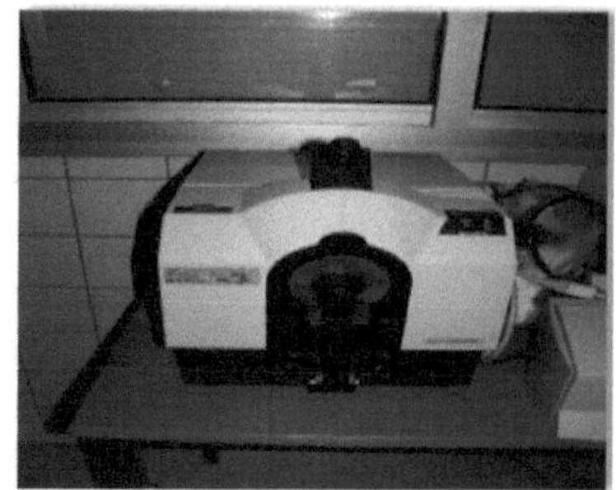

Figure 13: Spectrophotometer

a. Methods of colorimetric evaluation

The dyeing quality was assessed by calculating the absorbance (K/S) of the dyed samples. For this purpose, the reflectance of the dyed samples was measured at 420 nm using a Spectraflash colorimeter equipped with DataMaster 2.3 software, DataColor International (USA) with a D65 light/ 10 ° obseverator.

5.2. Method of checking the solidity of dyed material

a. Frictional strength (ISO 105-X12)

> **Material**

In order to measure the frictional strength, we made a crockmeter (figure 14), grey scale, a woolen fabric cut into squares of dimensions 5cm X5cm and two test pieces each of dimensions 14cmX5cm.

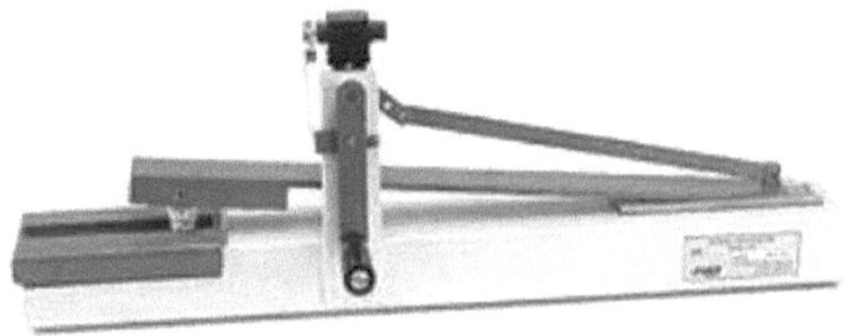

Figure 14: Crock meter

> **Method**

This test determines the resistance of dyes to washing with a detergent

❖   Dry friction

➤ Place the conditioned rubbing cloth flat on the peg end of the crock meter.

➤ Rub the dry test tube 20 times in a straight line back and forth; make a total of 10 back and forth movements, exerting a downward force y Remove the test tube and pack it.

➤ Remove all extraneous fibrous material that could interfere with the evaluation.

❖ Wet friction

➤ The same test was performed except that the control fabric was wetted with distilled water and wrung out.

➤ The samples were then dried at room temperature before the results were evaluated using the grey scale.

b. *Light fastness (NFG 07 067)*

➤ **Material**

In order to carry out the measurement of the light fastness a sunset (xenon lamp, figure 15) and a test tube of dimensions 4,5cmX1cm.

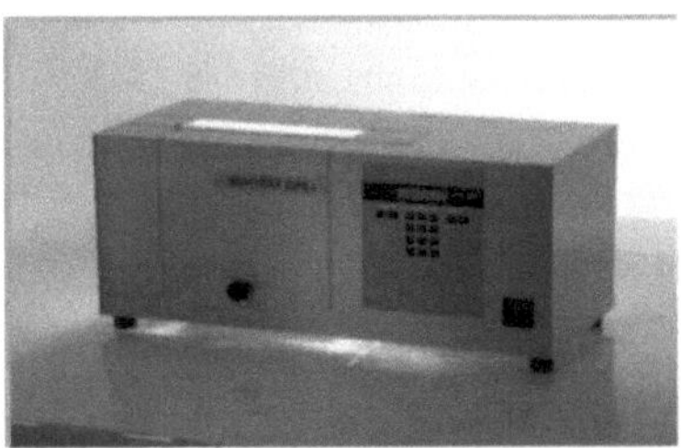

Figure 15: Sunset

➤ **Method**

This test is used to determine the resistance of dyes to light exposure.

➤ Display the test tubes and the set of standards, each of which is hidden to assess degradation.

Compare after exposure to light, the sample tested with the scale of blues treated under the same conditions.

Light fastnesses are rated from 1 to 8. For this type of fastness, we used a blue range (8 reference samples of blue-dyed wool) on which we evaluated the degradation of the specimens to be tested.

c. *Wash fastness (NFG 07-200)*

> **<u>Material</u>**

To carry out the test of solidity to the washing, we used for this fact autowash, detergent and control fabrics (multifibre band) and two test-tubes each of the dimensions 10cmX4cm.

> **<u>Method</u>**

This test is used to determine the resistance of dyes to washing with a detergent.

> Sew the test specimen to a multi-fibre fabric specimen of the same dimensions.

> Place the test tube in the autowash container containing : - 150 ml of distilled water.

> 4g/l of detergent.

Perform the wash test under the following conditions:

> Washing time=30 minutes.

> Temperature = 50°C.

> After washing, rinse with distilled water and then dry at a temperature not exceeding 60°C, placing it in such a way that the three parts are not in contact except by a seam along the short side.

The degradation of the test piece compared to the unwashed sample is assessed using the grey scale. Disgorgement is assessed by examining the tested and untested multi-fibre strip.

6. Determination of flavonoids

> **Material**

To carry out the determination of flavonoids, we used the spectrophotometer.

The spectrophotometer comprises a source of white light (tungsten lamp), a dispersive system for selecting the wavelength of the radiation and a detection system for measuring the light intensity of the monochromatic radiation passing through the solution. It compares the incident and transmitted light intensities and displays the absorbance (A) via an electronic circuit:

$$\mathbf{A = \varepsilon.\, 1.\, C}$$

With :

- A: the absorbance of the solution

- C: the concentration of the dye solution in mol/L

- l: thickness of the solution through which the light beam passes

- ε the molar absorption coefficient or molar extinction coefficient.

To validate the Beer-Lambert law it is necessary to work in monochromatic light, the solutions used must be diluted, homogeneous, and the solute must not give reactions under the effect of incident light (Zeghad Nadia, 2009).

➢ **Method**

The flavonoid content of our extracts is estimated by the AlCl3 method.

➢ 200 µl of each sample tested (1mg/ml)

➢ 300 µl sodium nitrite (5%)

➢ 500 µl AlCl3 (10%) [After 6 minutes].

➢ 1 ml NaOH [After 5 minutes]

➢ Incubate for 15 min at room temperature

➢ Read the absorbance at 510 nm, quercitrin is used as a standard, the amount of flavonoids is estimated in mg EQ/ml of dry plant extract.

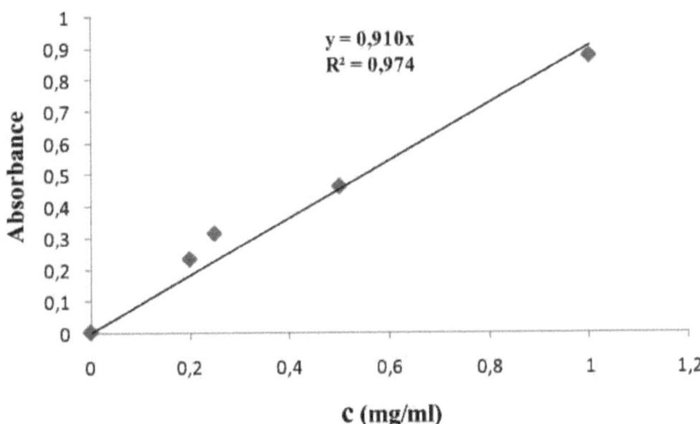

Figure 16: Calibration curve

Conclusion

This chapter has included all the material and the different methods and techniques used to conduct our experimental study.

Chapter III: Development of extraction and natural dyeing processes

Introduction

Throughout this chapter, we shall be interested in the development of a process for dyeing wool using the extract of *Asphodelus tenuifolius*. To do this, we will study the experimental parameters that affect the extraction process of this plant as well as the process of dyeing wool with the resulting extract. In addition, we will study the effect of different types of mordanting on the dyeing quality of the wool and its fastnesses obtained.

1. Development of the extraction process

In order to determine the effect of each of the following parameters: pH, temperature, time and mass of the plant material, on the dyeing quality of the wool, we will vary one of them each time while keeping the others fixed. We recall that the standard conditions set are:

- pH= 12

- Temperature= 80 °C

- Time= 60 min

- Mass of the plant = 1g

- Volume of water = 100 ml

1. Study of the effect of extraction pH

We start by studying the effect of the variation of the pH of the extraction on the flavonoid content as well as the degree of absorption (K/S) measured for the woolen fabrics dyed with the aqueous extract obtained.

Figures 17 and 18 show the evolution of the flavonoid content and the degree of absorption as a function of the pH variation of the extraction bath.

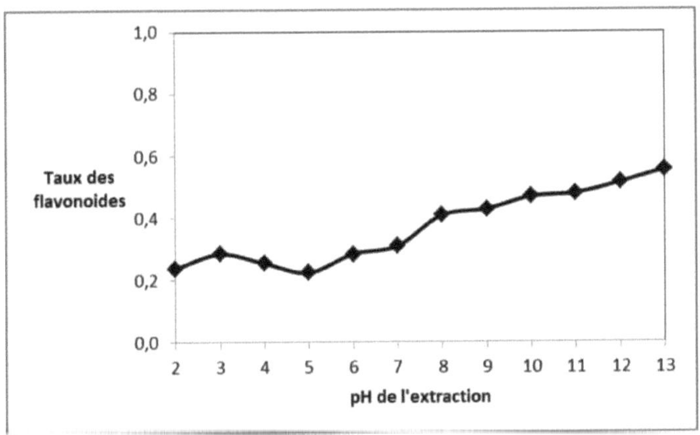

Figure 17: Evolution of flavonoid content as a function of extraction pH

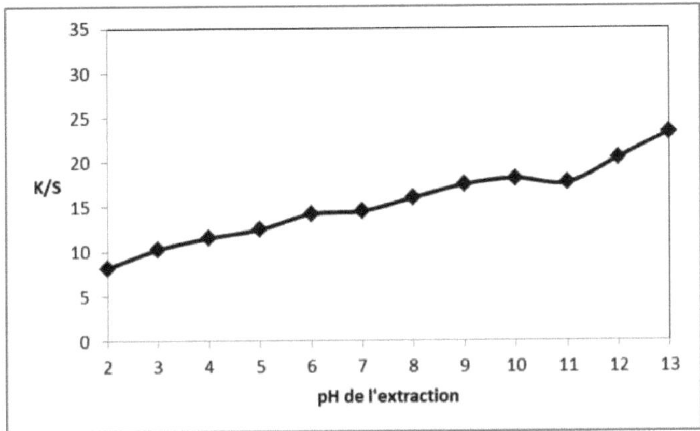

Figure 18: Evolution of K/S as a function of extraction pH

Referring to curve 17, we notice that the maximum of extracted flavonoids is obtained at basic pH values. This behavior can be attributed to the acidic character of these colored compounds (Ksibi et al., 2014). In addition, the outer walls of plant materials are usually made up of the cellulosic materials that gain anionic charges, which helps to create repulsive forces causing the easy rupture of these walls and the extraction of coloring materials (El-Nagar et al., 2005).

Similarly, from the results presented in Figure 18, we see that the degree of absorption (K/S) increases with the increase of the extraction pH until reaching the maximum value which is equal to 23.364 for a pH value equal to 13. Thus, we can see that the more the pH increases, the more the dye yield is important. From this we can conclude that the maximum of extracted coloured compounds is obtained for basic pH.

2.   Effect of extraction temperature

The effect of temperature on the extraction efficiency is studied since this experimental parameter has a great effect on the solubility and mass transfer rate in the extraction bath (Spingo et al., 2007).

Figures 19 and 20 show respectively the evolution of the flavonoid content and the degree of absorption (K/S) as a function of the extraction temperature.

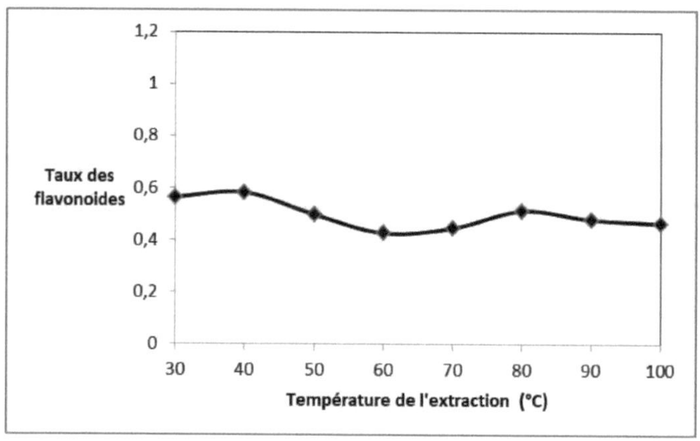

Figure 19: Evolution of flavonoid content as a function of extraction temperature

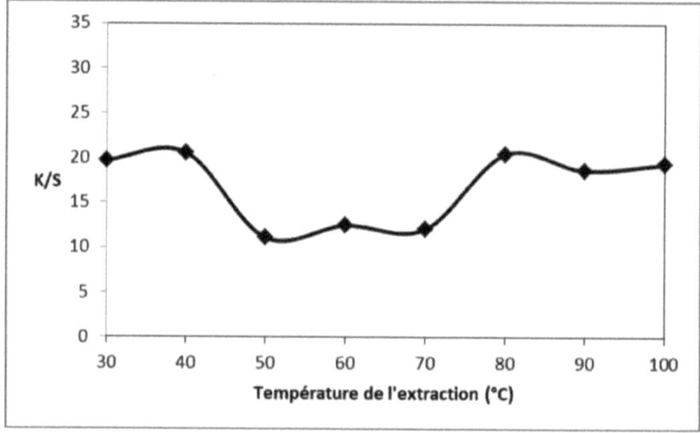

Figure 20: Variation of K/S as a function of extraction temperature

By observing the figure19, we can note that at low temperatures, we obtained a high rate of flavonoids. While, following the rise in temperature, the rate of flavonoids decreases.

This shows that temperature has a negative effect on these phenolic compounds. Indeed, the increase of this parameter can generate the decomposition of these molecules at high temperatures. This result was also reported by the studies of Laleh et al. (2006) and the studies of Türker and Erdogdu (2006).

Thermal degradation of flavonoids at high temperatures can cause hydrolysis of glycoside bonds as well as other bonds, leading to the formation of other molecules responsible for the dark brown colour (JU and Howard, 2003; Laleh et al., 2006). This explains well the increase in flavonoid content and consequently the dye yield from the temperature of 80°C as shown in Figure 20.

3.   Effect of extraction time

The time factor is a very important parameter for the good progress of the extraction process. For that, let us vary the duration of the extraction from 30 min to 120 min in order to evaluate the effect of this experimental parameter on the rate of flavonoids as well as on the degree of absorption (K/S) (see figures 21 and 22).

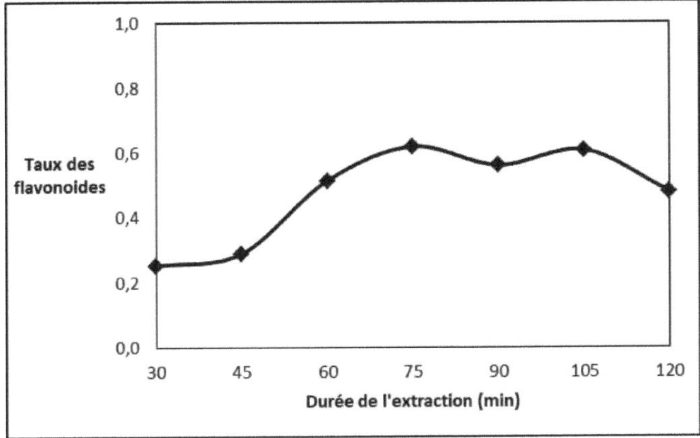

Figure 21: Evolution of flavonoid content as a function of extraction time

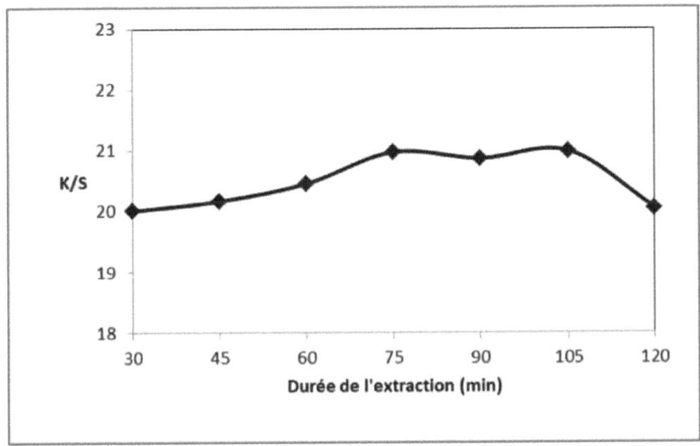

Figure 22: Evolution of K/S as a function of the extraction time

By observing these two figures 21 and 22, we observe that the flavonoid rate as well as the degree of K/S absorption increase with the increase of the duration of the extraction until reaching their maximums for a duration equal to 105min. Indeed, the longer the stirring time, the more we promote the extraction of phenolic compounds by promoting contact between the different particles of the bath and thus allowing the bursting of some cells that will release their contents into the liquid medium (Ben Amor, 2008) .

4.   Effect of plant mass

The influence of the mass of *Asphodelus tenuifolius* on the extraction is studied. We vary the plant mass from 2 g to 10 g while keeping the other operating conditions constant (a pH equal to 12, a temperature equal to 80°C, a duration equal to 60 min and a volume of water equal to 100ml).

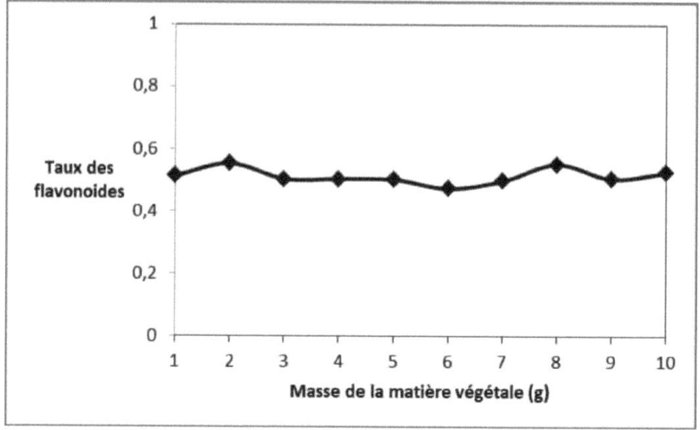

Figure 23: Evolution of flavonoid content as a function of mass of Asphodelus tenuifolius

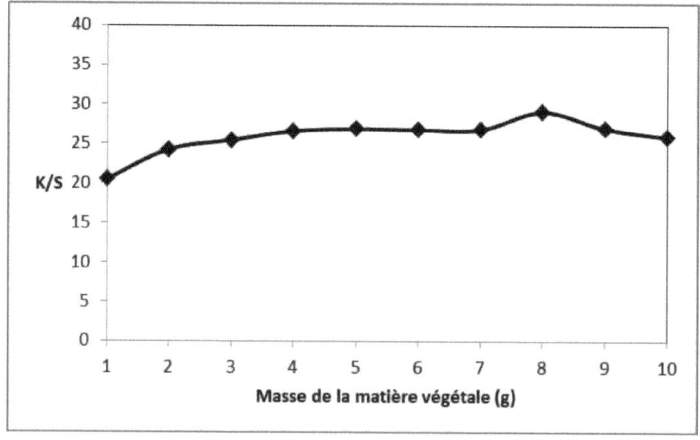

Figure 24: Evolution of K/S as a function of mass of Asphodelus tenuifolius

According to the results presented in figure 23, we can note that the maximum rate of flavonoïdes is obtained for a mass equal to 8g. Indeed, for masses included between 3g and 8g, we have almost a constant rate of flavonoïdes, which is also translated by the values of (K/S) presented by figure 24. We thus have a better degree of absorption for a mass of 8g of vegetable matter. By exceeding this mass, we note a decrease in the values of (K/S). This can be explained by the saturation of the wool fibre in terms of dye material under these experimental test conditions. Thus, any further amount of dye will be insoluble and will not contribute to a better dyeing yield.

II. Development of the dyeing process

After studying the effect of operating conditions on the extraction process of colorants from *Asphodelus tenuifolius*, we set the extraction conditions as follows: a pH equal to 12, a temperature equal to 40°C, a duration equal to 60 min and a plant mass equal to 4g.

We will now study the different experimental parameters acting on the quality of the dye obtained with the aqueous extract of the plant *Asphodelus tenuifolius*.

For this purpose, we will set the operating conditions for dyeing as follows: a pH of 3, an extraction temperature equal to 80°C and an extraction time of 60 minutes.

1. Effect of dye pH

We will begin by studying the effect of pH on the quality of the dye. For a temperature equal to 80°C and a time equal to 60°C, we varied the pH from 3 to 7. The results are summarised in figure 25.

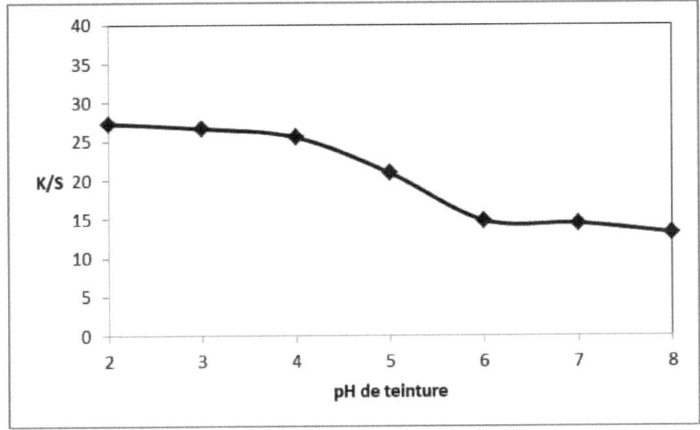

Figure 25: Evolution of K/S as a function of dyeing pH

From figure 25, we notice that the dyeing yield reaches a maximum value for a pH equal to 2 from which the yield obtained starts to decrease. In fact, the best dyeing yield is obtained in an acidic environment. This can be explained by the fact that the wool at an acid pH lower than 4.3 is positively charged thanks to the increase in the number of NH_3^+ (Chariat et al., 2005). This is accompanied by the swelling of the wool fiber due to these acidic pH accompanied by high temperature (80°C) and mechanical agitation. This will allow an increase in the diffusion of the dye inside the wool fibre. On the other hand at basic pH, there is dominance of (COO-) groups on the wool fiber (Monkholrattanasit et al., 2011).This will lead to electrostatic repulsion between the phenolic compounds causing a decrease in the degree of absorption (K/S) (Punrattanasin et al., 2013).

2.   Effect of dye temperature

In order to study the effect of temperature on the dyeing process, we varied this parameter from 30°C to 105°C, keeping the other parameters fixed (pH equal to 3 and duration equal to 60 min). The results obtained are shown in figure 26.

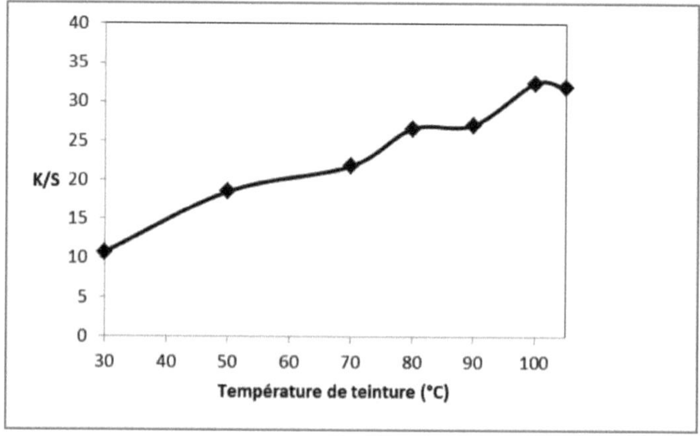

Figure 26: Evolution of K/S as a function of the dyeing temperature

Figure 26 clearly shows that the degree of absorption increases with the increase in temperature until reaching a maximum value of 32.44 for a temperature equal to 100°C. In fact, by increasing the temperature, we favour the migration of the dye molecules in the fibre, given the decrease in surface tension and the swelling of the fibre, and therefore we favour the mobility of the dye molecules and their migration in the fibre and consequently the diffusion inside the fibre (Chariot et al., 2005).

3.   Effect of dyeing duration

The time parameter has a great influence on the course of the textile fibre dyeing process. Indeed, a sufficient treatment time allows, on the one hand, to bring the fibre to its equilibrium state. On the other hand, the fibre reaches its saturation in terms of dyeing matter. Also, it ensures a better diffusion of these dye molecules inside the fibre.

Therefore, in order to emphasize the effect of this parameter, we varied the duration from 30 min to

120 min, keeping the other dyeing parameters constant (pH 3 and temperature 80°C). We obtain the results described in figure 27.

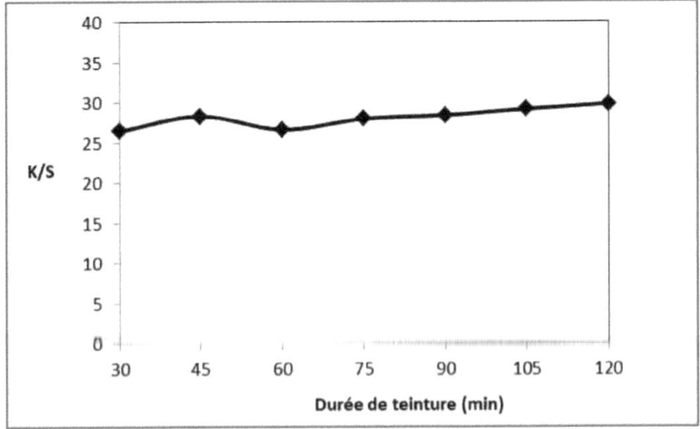

Figure 27: Evolution of K/S as a function of dyeing time

By observing figure 27, we can see that the degree of absorption (K/S) increases as the dyeing time increases. In fact, the time parameter conditions the good distribution of the dye and the maximum exhaustion of the dye bath. Therefore, the longer the dyeing time, the more dye we will have that is perfectly fixed on the fibre until equilibrium is reached.

III. Study of the effect of etching processes

1. Mordanting technique

Most of the dye molecules extracted from plants cannot form strong bonds with the textile fibre. Indeed, many of them can combine with various metal salts because their structure presents one or more chelation sites designating the association of a dye with a metal. The mordanting allows on the one hand to improve the quality of the solidities of the dyes and on the other hand to have other shades.

In this section, we studied the influence of mordanting on the dyeing quality of wool with the aqueous extract of *Asphodelus tenuifolius*.

The different metal salts that we have used as mordants are :

- ➤ Aluminium
- ➤ Iron II sulphate
- ➤ Copper sulphate
- ➤ Tin chloride

> Zinc chloride

> Tannic acid

The amount of mordant added is equal to 3 % of the mass of the sample. The mordant was added at different stages of the dyeing process:

❖ Dyeing process with prior mordanting

❖ Process for dyeing with simultaneous mordanting

❖ Dyeing process with subsequent mordanting

The dyeing quality was evaluated by measuring the degree of absorption (K/S), the colorimetric parameters (L*, a*, b*) as well as by the light fastness tests

2. Effect of etching on the degree of absorption (K/S) and the colorimetric coordinates (L*, a*, b*)

Table 3 shows the shades obtained as well as the measured absorption values and the values of the parameters (L*, a*, b*). These values correspond to samples dyed with and without mordanting.

Table 3: *Dyeing results with and without mordanting (shades obtained, (K/S) and colorimetric data (L*, a*, b*, C* and h*)*

Method	Bite	Samples	(K/S)	L*	a*	b*	C*	h*
	Without etching		26,62	41,08	7,88	27,76	28,85	74,14
	Alum		26,98	40,90	10,13	31,45	33,04	72,14
	Tin chloride		25,86	41,30	10,00	30,98	32,56	72,11
	Iron sulphate		28,22	34,96	6,25	25,29	26,06	76,12
Pre-ordering	Tannic acid		24,51	38,58	9,20	28,07	29,54	71,85
	Zinc chloride		26,96	37,28	8,95	27,74	29,15	72,12
	Copper sulphate		25,56	37,84	8,37	27,89	29,12	73,29
	Alum		25,13	43,22	10,25	33,48	35,01	72,98
Simultaneous etching	Tin chloride		26,31	42,54	10,96	32,90	34,68	71,58
	Iron sulphate		24,83	38,57	6,06	26,81	27,48	77,25
	Tannic acid		23,18	43,10	10,12	32,18	33,74	72,55
	Chloride		24,29	41,72	10,75	31,52	33,30	71,17
	Copper sulphate		24,56	40,79	8,10	29,96	31,03	74,87
	Alum		30,29	38,63	10,14	32,56	34,11	72,71
Post-mordanting	Tin chloride		26,15	40,87	9,97	31,00	32,57	72,17
	Iron sulphate		32,07	20,30	3,91	10,42	11,13	69,45
	Tannic acid		22,61	39,17	9,49	26,38	28,04	70,21
	Zinc chloride		31,42	33,18	9,76	25,82	27,61	69,30
	Copper sulphate		32,83	29,77	9,60	24,91	26,70	68,92

Firstly, from the re-dyeing tests, it was found that the (K/S) of wool fabrics dyed with *Asphodelus tenuifolius* extract are generally very high. This result could be attributed to the higher affinity of phenolic compounds with wool fibre. The colorimetric data of wool fabrics dyed by the different techniques in the presence of mordants are given in Table 3.

This table shows that in the case of wool dyeing using the pre-mordanting method, iron gave a higher dyeing yield (K/S =28.22) than the other mordants used. The degree of absorption (K/S) of wool fabrics increased in the following order:

Iron > Alum > Zinc chloride > Tin chloride > Copper sulfate > Tannic acid

For the simultaneous mordanting method, we found that Tin Chloride gave the highest dye yield (K/S= 26.31). However, we notice that Tannic acid decreased the dye yield. This may be due to the decrease in solubility of the dye substances when this metal salt was added to the dye bath. The degree of absorption (K/S) of the wool fabrics increased in the following order:

Tin chloride > Alum > Iron sulphate> Copper sulphate> Zinc chloride> Tannic acid

The best results were obtained with the post-etching method. This technique showed a higher intensity of colour compared to fabrics dyed using the other two methods.

This may be due to the great capacity of the metal ions to form complexes with the dye molecules in this technique. While in the case of the simultaneous mordanting method, some of the dye is lost due to the formation of an insoluble complex in the dye bath itself, and in the pre-etch process some of the mordant is stripped from the dye bath, which subsequently forms an insoluble complex with the dye molecules in solution.

Looking at Table 3, we can see that the best degree of absorption (K/S) is obtained with the copper sulfate mordant (K/S =32.83).

This increase in K/S values may be due to the ability of metal mordants to form coordination complexes between both the hydroxyl groups of the dye molecules and the functional groups of wool such as amino and carboxylic acid groups. The degree of absorption (K/S) of the wool fabrics increased in the following order:

Iron > Copper sulphate > Zinc chloride > Alum > Tin chloride > Tannic acid

3. Effect of mordanting on the strength properties of dyed fabrics

Mordants are used in the textile industry to improve dye fastnesses. Indeed, we have studied the influence of each mordant used on some dye fastnesses, namely: fastness to washing, to light and to rubbing.

Table 4 shows the results of the fastness tests carried out on the wool samples that were dyed with our process in the absence and presence of an additional mordanting step for the dyeing processes with pre-etch, simultaneous mordanting and post-etch for the different types of mordants studied.

The rubbing, washing and light fastness tests are carried out according to ISO105-X12, ISO105-C06 and ISO105-E04 respectively.

Table 4: Properties of the dye solidities with and without mordants

Dyeing method	Bite	Washing ISO 105C06	Light ISO 105-B02	Friction ISO 105-X12 Dry	Wet
Without etching	-	4	4	4	3-4
Pre-etching	*Alum*	4-5	5	4	4
	Tin chloride	4-5	5	4-5	4
	Iron sulphate	4-5	5	4-5	4-5
	Tannic acid	4-5	5	4	4-5
	Zinc chloride	4-5	5	4-5	4-5
	Copper sulphate	4-5	5	4-5	4-5
Mordanting simultaneous	*Alum*	4-5	5	4-5	4
	Tin chloride	4-5	5	4	4
	Iron sulphate	4-5	5	4-5	4
	Tannic acid	4-5	5	4	4
	Zinc chloride	4-5	5	4-5	4
	Copper sulphate	4-5	5	4-5	4
Subsequent mordanting	*Alum*	4-5	5	4-5	4
	Tin chloride	4-5	5	4-5	4-5
	Iron sulphate	5	7	5	4-5
	Tannic acid	4-5	6	5	4
	Zinc chloride	4-5	6	4-5	4-5
	Copper sulphate	5	6	5	4-5

Looking at Table 4, we notice that the rubbing, washing and light fastnesses are good. This result can be attributed to the high affinity of phenolic compounds of *Asphodelus tenuifolius* and the studied protein fiber.

In addition, we find that all three etching methods slightly improved the rubbing and washing fastnesses.

However, we observe a significant improvement in light fastness especially in the case of post-etching with iron sulfate. This can be explained by the ability of iron sulfate to form complexes between phenolic compounds and wool that are more stable to light.

Conclusion

This research has shown that the plant *Asphodelus tenuifolius* which is a plant with medicinal character, could be successfully exploited also as a promising source of natural dye for dyeing wool fibre.

The addition of metallic salts as a mordant has improved the solidity to washing, rubbing and light.

General conclusion

The objective of this work is to study the possibility of making a natural dyeing of wool from a plant with medicinal character: *Asphodelus tenuifolius*.

To achieve this we followed the following methodology:

- First, we conducted a literature review that allowed us to recall the dyeing properties of wool on the one hand and to know the dyeing potential of *Asphodelus tenuifolius on the* other hand. In this part, we also presented the different extraction and dyeing techniques that could be useful to us later.

- Secondly, we have presented the different products, materials and experimental protocols used in the study presented throughout this book.

- Thirdly, we presented the results of the experimental study carried out. In order to develop a dyeing process with the extract of *Asphodelus tenuifolius*. We started with the study of the influence of experimental effects on the extraction process of *Asphodelus tenuifolius*. Following the results found, we were able to determine the best conditions for the extraction, namely: pH of 12, a temperature equal to 40°C, a treatment time equal to 60 min and a mass of 4g. The evaluation of the flavonoid content of our extract confirmed the dyeing character of the plant material since these compounds present an important dyeing potential according to the previous works.

Then, in order to develop a protocol for dyeing wool with the aqueous extract of *Asphodelus tenuifolius*, we carried out a study of the effect of the experimental parameters on the dyeing process. Indeed, the optimal conditions for dyeing are: pH equal to 2, temperature equal to 100°C and a duration equal to 120 min.

Finally, we examined the influence of the application of the mordanting dyeing method on the dyeing quality as well as on the dyeing fastnesses of the wool. This study resulted in a slight improvement in the fastnesses to rubbing and washing, while the fastnesses to light were clearly improved. This improvement is more significant when using iron sulphate as a mordant in the post-mordanting with a value of 7 compared to 4 without mordanting.

Bibliographic references

Akroum ,S. , 2011, thesis Analytical and biological study of natural flavonoids.

Amiot-Carlin, M. J. 2004, ,Industrial extraction of bioactive substances from plants case of polyphenols. The SILAB's days.

Batanouny, K.H. 1999, Wild Medicinal Plants in Egypt. The Palm Press. Cairo.

Bellakhdar, J. , 1997, La pharmacopée marocainetraditionnelle. Ancient Arabic medicine and popular knowledge. IBIS Press.

Ben Amor, B. , 2008, thesis on the technological aptitude of the plant material in the extraction operations of active principles; texturing by controlled instantaneous relaxation DIC.

Brachet, A., S. Rudaz, et al. , 2001 "Optimisation of accelerated solvent extraction of cocaine and benzoylecgonine from coca leaves." Journal of Separation Science 24: 865- 873.

Chariat, M., Rattanaphani, ., Bremner,J.B and Rattanaphani, V., 2005, An adsorption and kinetic study of lac dyeing on silk, Dyes Pigments, 64, 231-241.

Christophe JACQUES, 2003, study of the valorisation of keratinic wastes by thermo-mechanical-chemical way in order to obtain continuous filaments: specific case of wool, thesis.

D. Grigonis, P.R. Venskutonis, B. Sivik, M.Sandahl and C.S Eskilsson, 2005, Comparison of different extraction techniques for isolation of antioxidants from sweet grass(Hierochloe odorata), The Journal of Supercritical Fluids, 33(3) 223-233

El-Nagar, K., Sanard, S.H, Mohamed, A.S.Ramadan, A., 2005, Mechanical properties and stability to light exposure for dyed Egyptian cotton fabrics with natural and synthetic dyes, Polym-Plast. Techno.Eng. 44,1269-1279.

H. Ghouilaa, N. Meksib,W. Haddarb, M.F. Mhennib, H.B. Janneta, 2011,Extraction, identification and dyeing studies of Isosalipurposide, a naturalchalcone dye from Acacia cyanophylla flowers on wool, sciencedirect.

Jocic D., Vilchez S., Topalovic T., Navarro A., Jovancic P., Julia M.R., Erra P.,2003, "Chitosan / acid dye interactions in wool dyeing system", Carbohydrate Polymers 60, 51-59.

Ju, Z.Y. and Howard, L.R ,2003, Effects of solvent and temperature on pressurized liquid extraction of anthocyanins and total phenolics from died red grape skin, J.Agric.Food chemistry, 51, 5207-5213.

Kaufmann, B. and P. Christen, 2002, "Recent extraction techniques for natural products: Microwave-assisted extraction and pressurized solvent extraction." Phytochemical Analysis 13: 105-113.

Laleh, G.H., Frydoonfar, H., Heidary, R., Jameei, R., and Zare , S., 2006, The effect of light, temperature, pH and species on stability of anthocyanin pigments in four Berberies species, Pak.J.Nutr., S,90-92.

L.Danielski, L.M.A.S. Campos, L.F.V.Bresciani, H.Hense, R. A.Yunes and S.R.S Ferrira, 2006,Marigold (Calendula officinalis L.) oleoresin: Solubility in SC-CO2 and comparison profile, Chemical Engineering and Processing, 46(2) 99-106

Meena Devi. V.N1 *, Ariharan . V.N 2 and Nagendra Prasad P2 . 2013, Anatto:Eco-friendly and potential source for natural dye. International research journal of pharmacy.

Monghhbrattanasit, R., Krystufek, J.,Wiener, J. and Vikova , M., 2011. Dyeing, fastness and UV-protection properties of silk and wool fabrics dyed with eucalyptus leaf extract by exhaustion process. Fibers Text. East. Eur. 19, 94-99.

PENCHEV Petko Ivanov , 2010, Study of extraction and purification processes of bioactive products from plants by coupling low and high pressure separative techniques ,thesis.

P.Kubelka, J., 1948,Opt.Soc.Am, 38, 448-457.

Punrattanasin, N., Nakpathom, M., Somboon , B., Narumol, N., Rungruangkitkrai, N. and Monghholrattanasit, R.2013. Silk fabric dyeing with natural dye from mangrove bark (Phizophoraa piculata Blume) extract, Ind.Crop.Prod.49, 122-129.

Mansour, R., 2005, Optimization of the dyeing of natural fibers with vegetable dyes, Master's thesis in textile engineering, ENIM, University of Monastir.

Sihvonen, M., E. Jarvenpaa, et al. 1999, "Advances in supercritical carbon dioxide technologies" Trends in Food Science and Technology 10: 217-222.

Spingo, G., Tramelli, L. and Feveri, D.,M., 2007, Effects of extraction time, temperature and solvent on concebtration and antioxidant activity of grape marc phenolics J.Food Eng., 81:200208

Tüeker, N. and Erdogdu, F., 2006, Effects on pH and temperature of extraction medium on effective diffusion coefficient of anthocyanin pigments of block carrot (Daucuc Carota Var. L.), J.Food.Eng, 76, 579-583

Zeghad, N., 2009, Etude de contenu polyphénolique de deux plantes médicinales d'intérêt économique (Thymus vulgaris, Rosmarinus officinalis) et évaluation de leur activité antibactérienne, thesis.

I want morebooks!

Buy your books fast and straightforward online - at one of world's fastest growing online book stores! Environmentally sound due to Print-on-Demand technologies.

Buy your books online at
www.morebooks.shop

Kaufen Sie Ihre Bücher schnell und unkompliziert online – auf einer der am schnellsten wachsenden Buchhandelsplattformen weltweit! Dank Print-On-Demand umwelt- und ressourcenschonend produziert.

Bücher schneller online kaufen
www.morebooks.shop

Printed by Books on Demand GmbH, Norderstedt / Germany